AF334072

FIREFIGHTERS
THE WORLD
OVER

Government Service the World Over Series

POLICEMEN THE WORLD OVER
POSTMEN THE WORLD OVER
FIREFIGHTERS THE WORLD OVER

This book is dedicated to
the professional and volunteer firemen
all over the world who have
bravely fought fire
through the ages.

FIREFIGHTERS THE WORLD OVER

Written and Illustrated by

Floyd James Torbert

HASTINGS HOUSE

Publishers • NEW YORK

Contents

Note: Countries are arranged in relative geographical position from west to east.

Copyright © 1967, by Floyd James Torbert
All rights reserved. No part of this book
may be reproduced without
written permission of the publisher.

Published simultaneously in Canada
by Saunders, of Toronto, Ltd., Don Mills, Ontario.
Library of Congress Catalog Card Number: 67-21242
Printed in the United States of America

FIRES
AND FIREFIGHTERS

Since time began, the peoples of the world have lived in fear of fires — fires caused by lightning, war, or man's own carelessness. But when fire threatened to destroy any part of his property, man fought back. He did not merely surrender. However, he has not always been the winner in this endless struggle.

For many, many centuries only the most crude methods were used to fight this cruel peril. Little more was done than to throw water onto the flames from buckets or pots. Yet as far back as the ancient days of Egypt, men realized that they must work *together* as a team against this powerful and ruthless enemy.

The oldest known device to aid them was a water-pump that was used in Alexandria about 200 B.C. A short time later, the Greeks and Romans invented hand-operated fire engines. However, with the fall of the great Roman Empire (around A.D. 330), these contraptions were soon forgotten.

In the Dark Ages that followed, it was believed that fires were vengeance wrought upon the people for their sins. So from the fourth to the 14th century, men often did not try very hard to put out blazes that flared up. For this reason, many families were made homeless and there was much needless suffering. It was not until the late 17th century that towns and cities started their own fire brigades.

In this period fire hazards were many. Homes were heated by burning wood or peat, and lighted by candle flame or by oil lamp. Cooking was done over an open hearth. As a result, sometimes the towns and villages and even cities were completely leveled once a fire broke out in a dwelling or in a shop.

The firefighters of old were not paid for their services and did not work full time. They all had other trades. It was not until about a hundred years ago that firemen were given money for their work.

Today modern equipment and machines help the trained firefighters do a much better job, although water is still the main weapon. No doubt we will always be threatened by this cruel menace, but the men of the fire departments all over the world will continue to lessen the dangers. They will find better ways of controlling and preventing all kinds of fires. They will still be ready to risk their lives for us twenty-four hours a day, every day of the year.

U.S.A.

In colonial Philadelphia, Benjamin Franklin formed one of the first fire companies in America. This company was made up of volunteers and he acted as its chief. Another famous American, George Washington, was a volunteer fireman for the Friendship Company of Alexandria, Virginia, before he became President of the United States.

Until about 1850 in the growing city of New York, large bells in watch towers were rung when a fire broke out. In those days the firefighters' biggest problem was how to get water quickly. The engines had to pump it from the nearest river, pond or well.

It was around this time that steam pumps took the place of the old hand pumps. These machines were able to pump a steady flow of water onto the blaze long after the firemen would have been exhausted.

At first the steam engines (pumps) were pulled by the firemen themselves, but soon horses took over. Some motorized fire pumpers were on the scene about 1870, although

horse-drawn engines stayed around for many more years. By 1910, gasoline-driven fire engines began to be manufactured for city fire companies. The last of the old faithful fire horses was finally put to pasture by 1928.

All the fire departments in the United States are run by the local city or township governments. There are now over a thousand of these with full-time members and about 14,000 volunteer fire companies.

Most large cities have "colleges" or schools to train men in the latest fire-fighting methods. One of the very first "fire drill schools" was opened in New York City in 1883.

The largest fire department in the world is the one in New York City. The powerful new engines include a SUPERPUMPER, SUPERTENDER and SATELLITE — three separate units of a single pumping system. Its WATER TOWER TRUCK is able to reach most of the many skyscrapers and the world's busiest harbor is protected by a large fleet of fireboats. Over 13,000 "blueshirts" are trained like an army to meet any kind of emergency. There are also nearly 2,000 auxiliary members available for duty. These men answer about 140,000 calls every year, but some 30,000 of these prove to be false alarms.

From 1608, when the entire Jamestown settlement in Virginia burned to the ground, until the present time, the American people have had disasters from fire. The biggest one is perhaps the least known. In 1871, Peshtigo was a prosperous lumber town in Wisconsin. A nearby forest fire engulfed it, trapping everyone there. Before it was over, 1,200 lives were lost. It happened on the same day as the inferno that ruined much of the city of Chicago. This, the most renowned of all fires in America, is believed to have started in a stable by a cow kicking over a lantern. It spread so furiously that no one was able to cope with it.

Today the firemen in the United States are ever ready to jump into action with the best know-how and the finest equipment of all time. Incidentally, the firehouse in Mount Holly, N. J. is said to be the oldest that is still standing and its Relief Fire Company No. 1 claims to be the oldest active volunteer fire company in the United States. It was organized in 1752!

CANADA
29

One of the leading industries of Canada is lumbering. There are millions of trees in this rugged land, but they have been hit hard by many fires. The Province of New Brunswick had the worst one of all time in 1825. Men, women and children joined the volunteer brigades to fight it, but theirs was a losing battle. A tract of four million acres of trees and farms had been severely ravaged before the fire burned out. Many towns were gobbled up in its hundred-mile path.

At one time the Canadian fire companies did not even own the horses that pulled the engines. They were hired by the town or city from nearby livery stables. The first paid fire department was formed in Toronto by fifty-six hearty men. This was in 1874. It was not until 1911 that motor trucks were used.

DECHERD PUBLIC SCHOOL
LIBRARY 7183
Decherd, Tennessee

In 1917 a horrible fire took place in Halifax, Nova Scotia. It was caused by the explosion of a munitions ship in the harbor and set fire to all sections of the city. Before it was brought under control the greater part of Halifax was gone. Some 1,500 persons lost their lives, including many firemen.

The long, cold winters make fire fighting in the far north a very perilous task. Many times the fire engines race over icy roads or through blinding snowstorms. When they do arrive, they often find that sub-freezing temperatures have shut off the water supply.

In spite of these hazardous conditions, the fire companies of Canada have a very fine record over the years.

GREAT BRITAIN

For many centuries, England was at the mercy of flames. Time and time again its cities were laid to waste. Homes were built so close together that even a small blaze would soon be out of hand. In the year 1666, almost all of London burned down. This terrible fire started when a baker spilled hot coals on the floor of his shop. The conflagration lasted for six days.

The people of that time realized something had to be done. They formed a few volunteer fire companies, but all the equipment they had consisted of pick-axes, ladders and leather water buckets. However, this small beginning was the turning point in fire-fighting history!

A short time later, fire insurance companies formed their own brigades. FIRE MARKS (small cast iron plaques) were nailed to the front of insured buildings. Each insurance company had a different fire mark so that if a fire broke out it knew where to send its firefighters. These men tried to save the houses that were insured by their firms.

The English invented a ladder on two wheels but many

years passed before the first steam pump came along. Even with these inventions the country was still plagued with big fires.

In more recent times, British fire brigades have been put to almost impossible tests during air raids. There were few in the First World War, but in World War II over 50,000 fires were caused by bombings. One blitz raid alone started about 2,000 fires in greater London. Even women manned the pumps during that crisis when all were firefighters. Many members of the fire services lost their lives while on active duty.

At present, London has a very large, modern fire department. It also has a fine fireboat fleet. All rookie firemen are given a three-months course of intensive training before they go on active duty.

The British Internal Ministry maintains a nation-wide fire inspection service. Factories, warehouses, docks and public buildings are visited nightly and on weekends by roving fire patrols. Automatic alarms and spraying systems have been installed in many places. The Ministry's Fire Institute develops fire protection techniques.

Many private businesses and factories have their own fire companies, but there are only a few volunteer units in Great Britain today.

Nearly all calls for aid are made by telephone instead of by street alarm systems like those in the United States. No longer are British cities in grave danger of being ruined by fire.

DENMARK

The Danes saw a need for firefighters over three hundred years ago. In the seventh century men were drafted for state and village fire brigades.

The worst fire they ever had to combat was in 1738. The brave action by these old-timers could not save the city of Copenhagen. After four horrible days, little was left but ashes.

For many years the Danish Fire Defenses were under the king's command. They were not very strong. It was not until the Royal Palace burned in 1884 that they were greatly improved. The loss of life from this fire was small but the disaster shocked the whole nation. About four years later, Copenhagen had a well-run, professional fire force.

Around 1900 a young Dane named Sofus Falck hit upon an idea that would help save lives. He wanted the RED CROSS SOCIETY of Denmark to work side by side with the fire brigades. Together they would give aid to the injured victims. This was done in small towns as well as large cities.

During World War II Denmark had quite a few bad fires from bombings and sabotage. Many lives were lost. For five long years of German occupation the Danish firemen did a heroic job. Tales still are told about the fire at the Gestapo Headquarters in Shellhuset. About two hundred Nazi soldiers and Danish traitors perished. By some miracle, most of the Danish prisoners managed to escape this inferno.

Today Denmark has two world-renowned SALVAGE CORPS. They are always at the scene of a fire, flood or any other disaster. One of their most important missions is to take any injured persons from the small Danish islands to hospitals on the mainland. They do this by airplane. In fact, all the firefighters of this country, along with their helpers, are well prepared.

The northernmost fire department in the world is located at the Thule* Air Base. This scientific outpost and radar site, with a population of 3,000 American industrial and military personnel and Danish servicemen, is 600 miles above the Arctic Circle. It has some of the most modern, efficient and specialized fire-fighting apparatus in the world. Fortunately, it has seldom had to be used, and then only for small electrical flareups and trash fires. This is due to the excellent fire prevention program, for here in the Arctic winds are strong and water freezes almost instantly.

*See Pronunciation Guide to Foreign Words whenever this symbol occurs.

GERMANY

In the early 1800's a German by the name of Carl Metz wanted a strong fire department for his country. Many other countries had them. When a fire swept Hamburg in 1842, most Germans agreed with him. After much talking and planning his dream finally came true.

By 1851 the Berlin Fire Brigades were keeping blazes from spreading or getting out of control. In the days when speed could not be taken for granted, these units were usually at the scene within minutes after being called. Hanburg, Hanover and other cities soon had similar brigades.

Fire control in Germany has been noted for its thoroughness. Its firefighters also have always been eager to try new ways. In 1929 they experimented with soap suds. The soap was placed at the top of a steam boiler. It was whirled about by water and forced through a hose. It smothered most fires faster than plain water. Three years later, the first "powder" fire engine in the world was tried out in Frankfort-on-Main. This apparatus squirted powdered natron, a chemical that did far less damage to buildings and their contents than water.

Germany probably has the most unusual firefighters of all. They are the nuns of Saint Joseph's Convent in Ursberg. Every task at the convent must be done by these women. Rules of their order forbid any help from men or boys. If a fire is discovered they must take care of it by themselves. They have many drills. The seventy nuns wear fireproof habits, smoke masks and leather hoods. There has been only one really bad fire and this was in their home for mental patients. They handled it like the expert firefighters they are.

In all the small towns of the Federal Republic of Germany, the firemen are volunteers. In the cities they are professionals. This is about the same as in the United States.

There are a few hard-to-get-at places in this nation. For many years, there have been fire patrols for the high mountain towns and villages. In the winter, they wear skiis.

Most of the biggest fires in Germany happened long before there were any regular firemen around. However, the two worst ones came during World War II. The first one hit Hamburg in 1943 and the other took place in Dresden two years later. Both were started by bombs dropped by Britain's Royal Air Force. In each case flames from the burning city were caught in big drafts, causing terrible fire storms. Thousands of people perished in the infernos and from the heat caused by them. Even those who jumped into rivers did not escape being burned.

At one time schools in Germany had special courses to teach the boys and girls the dangers of fire and its prevention.

In East Berlin today the firemen are under the command of the Army-Police Force. Beyond the wall in West Berlin, the firemen work for the city government.

THE NETHERLANDS

Hardened soldiers and convicts were the first known fire-fighters in Holland. Later on, about the year 1500, the whole population of the city of Amsterdam had to take part in fire-fighting programs. This was a royal order and a good one. As soon as the warning was sounded, the Dutch household became a beehive of activity. Every bucket in each home was quickly filled with water and placed on the front doorstep. Strong young men would pick up two at a time and race to the burning building. Others, already there, would douse the flames with the water from the pails.

Amsterdam and other cities in Holland had only volunteers until 1874 when the first professional fire brigade was founded.

The Netherlands has not had any great fire disasters in the past or in recent times. Even in World War II bombings and other enemy action did not cause wide-spread damage. The people and firefighters of this low country may well be proud of their fine record.

FRANCE

In the 13th century, Paris had its FIRE WATCHERS. Their job was to spot fires and then to arouse the people. It was *not* their duty to help put out the fires.

Then, four centuries later, there was an ancient order of monks who pitched in when needed. They took care of many fires, big and small. These brave men would hurry to the burning area and in their bare feet would climb ladders to the roof tops, carrying iron buckets filled with water. Many times they were the unsung heroes of their day.

France had no organized fire-fighting system until the early 18th century. In 1716 a fire force of 40 men was formed in Paris.

The old-time French firemen wore very colorful uniforms with brass helmets. They carried their crude apparatus to fires and also onto the battlefield in time of war. At first they were under the command of the Prefect (Chief) of

Police but in 1811 they were made part of the military corps. Fifty-six years after that they became a regular regiment, commanded by a colonel of the French Army.

In the First World War, the FIRE REGIMENT stayed in the burning cities until the last moment when they would be called up to the front lines for battle duty.

Over 2,000 fire bombs were dropped on the city of Paris during World War II. A regiment of 6,000 men bravely fought the many blazes and managed to keep the damage as low as possible under those terrible conditions.

The soldier-firemen of France, known as SAPEURS-POMPIERS*, have a reputation of being among the finest firefighters in the world.

ITALY

Rome was the first city in the world to have a regular fire department. In the time of Augustus Caesar (27 B.C. to A.D. 14) there were between six and seven thousand firefighters ready to protect the 14 regions of that ancient metropolis. Slaves, too, were often trained to take part.

Although Rome under the Caesars was a rich and beautiful city, it did have a slum section. Here the houses were built of wood. Each family kept a flame burning on an open altar at all times to honor the gods. It was no wonder that many fires resulted.

NOCTURNES* were stationed along the streets of Rome. They were men who would shout the alarm along the line to the nearest CASTRA*, or fire house. The fire company would then rush to the burning building. It was led by a CENTURION*.

These early Roman firemen wore metal helmets, leather jackets and leather trousers. Some would be carrying SIPHONES*, which were wooden hand-pumps. Others would be armed with axes, saws, hammers and iron bars. They used short ladders that could be hooked together for climbing to the tops of the highest structures.

The first squads to arrive were the AQUARI*. Each man carried a light earthenware vase. They would at once form a chain to the nearest cistern. These were wells filled with water from the great aqueducts leading into the city from rivers. Soon the jars would begin emptying a steady stream

into the hand-pumps. About this time a chariot would arrive bearing the PREFECTUS VIGILUM,* the fire chief, and he would take charge.

There were also "surgeons" or doctors ready to care for the injured or for those overcome by smoke. Pillow bearers were on hand, too. Each brought a large leather pillow stuffed with feathers in case anyone should be forced to leap out of a window.

The most feared and respected of all the Roman officials would be the last to arrive. He was the QUESTIONARIUS*, whose duties were like those of our present-day fire marshal. He was supposed to find out just how the fire started and who was to blame.

In the year A.D. 64, it is believed the tyrant ruler, Nero, set fire to Rome. The fire burned for eight days and nearly all the great city was a charred ruin.

The Roman fire-fighting setup remained until the fall of the empire in the fourth century. Although there were a few FIRE GUARDS after that, it was not until about 1672 that another regular force appeared once more — THE VATICAN FIREFIGHTERS.

Many years passed before several Italian cities began to form the same kind of units. Milan came along with a brigade of 83 men who had been excused from military duty. The uniforms worn by the early firemen of Italy were very fancy and colorful.

By the 1800's most of the nation was well protected, yet southern Italy still did not have fire services. It depended on the Italian navy for help.

When Mussolini became dictator in 1922, all the industrial cities and towns got modern equipment for their fire brigades. They soon learned the newest and best ways to put out fires.

One Italian city has always had problems even though it had plenty of water around and in it. That, of course, is Venice. It was not easy to get to a fire by way of the canals. For many years the firefighters used the slow-moving gondolas. Today a modern fleet of fireboats answers an alarm within minutes.

United States has less than four times the number of people, but it has ninety times more firemen than Italy. However, there are far fewer fires in this older country because so many of its houses and buildings are made of stone. Narrow streets do not slow down the Italian fire companies as their trucks are rather small in size.

When called up for the draft, the young men of Italy may serve their eighteen months in the Firefighting Corps. Rome has one of the world's finest and largest schools and training grounds for firemen.

Italian law requires that a member of the fire department be present at every movie and theater performance. The fireman is paid a little extra for this service.

As in Russia and several other European countries, there are no fire-alarm boxes on the street corners of Italy. Firemen are usually summoned by telephone.

TURKEY

Hundreds of years ago high fire towers were built in Turkey. The crowded cities were watched over by day and by night for sign of smoke or flame.

If anything was seen, the watcher would run from the tower to the nearest fire station. Sometimes this was quite far away. Other men would then go down the streets shouting warnings. They would tell the people in what part of the city the fire had started. Relatives and friends of anyone living near the blaze would then help fight it. Mostly they beat out the flames with coats and blankets. For many years this was the only way the Turks had to protect their lives and properties.

Turkey formed its first fire company about 250 years ago. For a long time it was part of the Turkish Army.

In 1865 the port town of Smyrna had a FIRE INSURANCE BRIGADE of 100 men. It used a small "fire wagon" that the men pushed themselves. Later on this brigade became the proud owners of a steam engine pulled by horses. Soon Ankara and Constantinople hired regular firefighters. They bought equipment from Italy.

Water had to be pumped from town fountains or wells, and since this took so much time, fire could spread at will. To make matters worse, houses were built close together.

The people of Turkey have seen many of their cities in smoldering ruins. Turkish firemen have battled some of the worst fires on this earth.

Constantinople, now called Istanbul, has had more big fires than any other city in the entire world. From 1750 to as late as 1954 it has been almost completely destroyed many, many times. One of the biggest destructive fires in all history took place there in 1756.

Now Turkey has plenty of fire hydrants. The old fire towers still are used, but present day Turkish firefighters are well trained in the newest methods. They keep fires from doing the great damage of the past.

GREECE

In Greek mythology the only ugly immortal was Hephaestus*, the God of Fire. The Greeks were awed by fire, as were all ancient peoples, but they tried to tame it. Like that of the Romans, their fire-fighting knowledge was forgotten during the Dark Ages.

Way back in 1400 B.C. the first large-scale fire in history took place on Crete, an island off the mainland of Greece. It was here that Knossus*, the greatest metropolis of the time, was burned down. We also know that Carthage, the first city in the world to have a million inhabitants, was burned to the ground by the invading Roman armies in the year 146 B.C.

Through the long centuries that followed, their "Fire God" has struck at the Greek people many times. Fortunately, the blows have not been as heavy as those suffered by other European countries. This is strange as the Greeks have not been well prepared to combat the dreaded foe.

When the royal castles at Athens caught fire in 1908, every one joined in fighting the blaze. Men, women and children came running with buckets of water carried by hand and on poles. It was a losing battle for the willing but unskilled firefighters.

For years this was known as "the land without a fire department." There were no brigades at all, except for a small military corps detailed to take care of any fires near their base.

Finally, in 1931, the government created the Fire Department of Greece with headquarters in Athens. It is run along military lines and its duties take in rescue and relief work during floods or earthquakes. In the time of war it becomes a civil defense force.

RUSSIA

Before 1800 Russia had no fire department. If a blaze started in one of its cities the soldiers stationed nearby were called to put it out. In villages and on farms, the townspeople and farmers had to fight fires the best they could by themselves.

The first uniformed fire company was formed in St. Petersburg (now Leningrad) in 1801. It was made up of disabled veterans from the army. By 1830 observation towers similar to those in Japan and Turkey were built. Wagons filled with large barrels of water were ready at all times.

In those days firemen lived in barracks with their families. As they were paid only when they were needed to fight fires, all of them had other jobs on the side, like shoemaking or carpentering.

During the severe Russian winters, large sledges or sleighs, drawn by three horses, were used to carry the firemen to the burning buildings. Other sledges brought the hoses and tools. Still others came with huge casks of water. In this way, there was no waiting to open the icebound water supply when they reached the fire.

As all its buildings were made of wood and even its streets were paved with pine logs, the old city of Moscow had several great fires. The first one was in 1237 when it was burned by the invading Mongols. In the 15th century it was set ablaze by the Tartars. This action trapped over 200,000 Muscovites, who were burned to death. This was one of the worst tragedies in Russian history. For some reason there were no other big fires until 1752 when thousands of Moscow homes were turned into ashes. Finally, war was the cause of another disaster — the Great Fire of 1812. This happened when the people of Moscow fired their own city in order to drive out the armies of the French conqueror, Napoleon.

The Russians have come a long way since those hard days. Now there are strict fire prevention laws and the firemen have very good equipment. Their fire engines have air horns like train whistles for warning signals.

The old watchtowers have been torn down but Russia now has lookout cabins atop its highest buildings. The men assigned to them have a rather easy job as there are very few fires to report. Indeed, the capital city of Moscow has not had a sensational fire for over thirty years.

To protect the vast oil fields near the Caspian Sea, there are FIRE TRAINS. They speed to any oil blaze as soon as it is discovered and spray it with powerful chemicals.

In the Soviet Union today firemen go through rigid training programs. They also have sports and exercises. It is a tough life for these men but the government frequently gives them awards for "courageous acts while performing their duties."

Unlike other cities of the world, Moscow has a Volunteer Firemen's Association. It has thousands of members, many of whom are young boys. The fire department, assisted by these volunteers, keeps Moscow's fire rating low — one of the lowest anywhere. Fire, once the terror of Russia, has finally been tamed.

EGYPT

In countries as crowded as Egypt, ways to prevent and sub-due fire have always been of great concern. The Ptolemies,* a Greek dynasty that ruled the land three hundred years before Christ, had some form of fire fighting but little is known about it. History tells us that during this time a water-pump was invented by Hiron, the Egyptian. This really was the start of all fire-fighting equipment.

Under the Fatimites,* Mohammedan rulers from 909 to A.D. 1171, the earliest known "Be Prepared for Fire" program existed. In the cities a squad of water-carriers with leather bottles would call each night at the central police station. Each water-carrier was followed by ten demolition laborers, who were armed with pick-axes. Moreover, all the merchants had to put big jars of water in front of their shops every night after closing time.

Much later on water was supplied by horse-drawn carts loaded with large barrels. Water-taps were not installed in Cairo until about 1875. These were a vast help to the fire-fighters.

Perhaps because the Egyptians began early to think about fire-fighting, their damage from fires has not been great compared with other countries. Indeed, Egypt has a right to be proud of its fire brigades, which have copied those of far-away London.

The Cairo Fire Brigade is one of the biggest and best trained in the world. It looks strange when this very modern company practices hose pipe drills in the very shadow of the centuries-old pyramids.

INDIA

India has had fire brigades for many centuries, but up to a hundred years ago they were made up of volunteers armed with nothing more than buckets of water or sand, and were of little help to anyone.

Many outbreaks snuffed out the lives of thousands of humans and animals. Except in the large industrial centers of Bombay, Delhi, Madras and Hyderabad, there were hardly any fire services worthy of the name. Those in other places were manned by untrained men with ancient equipment. Many parts of the country had no protection at all.

Then, too, India's poor water supply and bad roads made fire-fighting difficult. Conditions improved during the British rule in the 1800's, although India had only volunteer units until 1867. After that, more and more fire services were started and equipment was sent from England. Until 1921 the police department ran it, but many problems had to be solved when British control ended in 1947.

The new government of India decided to tackle the weak spots one at a time. A committee of fire experts was formed in 1950 to make plans. Soon the young men of India were being trained for fire-fighting duty. By 1956 the National Fire Service College opened at Rampur.

There is now an annual Sports Day in Delhi with many prizes for the firemen and their families. During Fire Prevention Week the fire brigades put on demonstrations of their skills.

Today, Delhi, India's capital and one of the oldest cities in the world, has a very modern fire department. So have nearly all the other cities in this ancient, populous land.

A familiar sight at fires in India is the presence of pigtailed girls wearing long white blouses and white pantaloons. They are a section of the Indian Home Guard and are trained to assist the firefighters and to meet all civic emergencies.

CHINA
15-03979

Fire has always played an important role in the lives of the Chinese people. In their early days they worshipped a God of Fire whom they called Chu Jung.* Every year noisy Fire Festivals were held for the fierce FIRE DRAGON. They tried to "bar the power of fire" by beating drums, ringing bells and waving colorful banners.

About a hundred years ago, the British came along with a more effective way to protect their colony in Shanghai. An INTERNATIONAL FIRE BRIGADE was organized. Its members were of many nationalities but they worked well together under British supervision. The best fire-fighting apparatus of that time was secured. This brigade soon became a model for the whole of Asia.

In 1901 the Chinese Fire-Fighting Forces were started as part of the police system. But the old method of throwing buckets of water on the flames still was used in most of China until recently.

Great quantities of fireworks are needed in China for the many celebrations and festivals. In the past some of the factories that made them blew up and big fires started from the sparks. A close watch is now kept over all these places, but they are still a hazard to the Chinese.

It is the law in China that huge tanks of water be set up at various locations in towns and cities. These are for the firemen to use any time a fire breaks out.

There are many assistant firemen in this over-populated land. They take care of the less dangerous jobs at the scene of the fire, such as directing traffic and holding back the crowds. Orders are given to them by the police.

Today most Chinese fire companies have gone modern, with up-to-date tools and machines. They also own some of the best fireboats in Asia. These are badly needed because thousands of people live in houseboats. There could be a major catastrophe if they were not well protected, because crude cooking methods have caused many of these wooden sampans to flare up suddenly.

One of the worst fires China ever had happened on the waterfront. This was in Chungking in 1949 when some 1,700 people were killed as panic made the firemen's efforts almost futile. Half the city was destroyed.

JAPAN

Japan has one of the oldest fire-fighting systems in the world, but the Japanese once believed fires were caused by evil fire demons. To ward off their visits, the people would try to frighten away these demons with colorful banners and lanterns. Neither this custom nor the old-time fire-fighters seemed to do much good as there were many, many fires.

Most of the houses and temples on this island empire were built of bamboo, paper and wood. The Japanese thought filmsy materials would do less harm to them if there were earthquakes, which were feared even more than fire. But because the buildings were not made of brick or stone, entire towns and villages were lost to the flames.

So great was this threat that sometimes a man was blamed if his home caught on fire. In the eyes of his fellow villagers, he was worse than a criminal. So he was beheaded.

In 1560 a hundred or more years before most European countries had fire-fighting services, there was one in Edo (the old name for Tokyo). It belonged to the mighty *daimyo,** the feudal lords. Fire watchers were stationed around the Mikado's palace. They were called the SAMU-

RAI* FIRE BRIGADE. However, there were no brigades to safeguard the lives or property of the common people.

The Samurai warden would look for the fires from a high wooden tower. When he saw something burning, he would run to a nearby firehouse. If the firemen were asleep (they slept with their heads resting on a bamboo pole) he would arouse them by hitting the pole from under their heads with a wooden mallet. This meant that none of them would have the excuse of having slept through the alarm.

It took another hundred and sixty years before there were fire companies to protect the Mikado's subjects. These brigades were known as the MACHINIKESHI.* They were made up of volunteers who had to do the best they could with bamboo ladders, hook-poles, ropes and hand pails.

The usual method in those days was to pull down the burning building to keep the fire from spreading. Some years later crude hand-pumps sprayed water but were of little help.

Japan has had many reasons to fear fires and earthquakes. It has suffered two of the greatest disasters in the history of man. The Great Fire of Meireki in 1657 leveled many towns and took the lives of 108,000 men, women and children. Then in 1923, after the Great Kanto Earthquake, a raging fire destroyed about 366,000 homes. Nearly 100,000 persons were killed before it died out.

The danger of fire in the cities is now lessened because of the many modern, fireproofed structures of steel and concrete. But there are still crowded streets of wooden homes and shops.

No place in Tokyo, one of the world's largest cities, is further than five minutes from a fire station. Tokyo also has a modern fleet of fireboats. No other port or harbor in the world has as many.

The Japanese firemen of today still combine the old with the new. The annual ceremony to review the fire-fighting forces is called the Shobo Dezome-Shiki.* It takes place each New Year's Day at the Imperial Palace Plaza. The firemen wear the traditional kimonos and perform stunts with colorful flags atop high bamboo ladders.

THE PHILIPPINES

For over three hundred years the Philippine Islands were held by Spain. The Spaniards cared little about fire problems there. The United States took the islands in 1898 during the Spanish-American War, after Admiral George Dewey defeated the Spanish fleet in Manila Bay.

Until 1910 there were but a few trained firemen and not nearly enough equipment. Manila, the largest city, had no separate fire department. Its small forces had been under the Department of Streets, Parks, Fire and Sanitation.

At this time some new steam engines and a few motor trucks were added to the firemen's old horse-drawn wagons. Slowly Manila's fire company began to shape itself into a first-class organization.

From 1898 to 1935, when the fire services were run by the United States, the fire chiefs always were Americans. In 1935 the Commonwealth of the Philippines was formed and native-born Filipinos were put in charge.

The war in the Pacific found the Japanese occupying the islands. Most of the modern fire apparatus was commandeered by the Nipponese Army. Many native firemen were taken for guerilla fighting against the invaders.

When the Japanese were driven out in 1945 by the Americans, a very weak fire department was left. The United States Army had to supply it with used army trucks, hoses, axes and ladders to make it effective once more.

About 4:15 in the afternoon of February 21, 1964, a fire broke out in a machine shop where welders were working. It spread rapidly in the downtown section of Manila, even though every city and suburban fire-fighting unit responded to the alarms. Naval and air force fire departments were called to help, too. The blaze continued all through the night and into the morning, devouring ten blocks of buildings. It was, by far, the worst fire to hit the islands since the battle for liberation nearly twenty years before.

CHILE

Chile can boast of the oldest fire department in South America. In 1863 the newly formed outfit in Santiago was put to a real test. The Church of the Campania caught fire during religious services. In the panic that followed 2,000 lives were lost.

The skillful firefighters of this country are *all* volunteers. Even the fire brigades of Santiago, with its population of over two million, serve without pay of any kind. Besides the native Chilean firemen, there are Spaniards, Englishmen, Yugoslavs, Germans and Frenchmen. All work side by side when the alarm sounds. Until about twenty-five years ago each wore the uniform of his own native country.

There have been some big problems for these firemen to tackle. Since most buildings in Chile are made of wood, a fire has always been hard to control. Then there have been many fast-spreading brush fires in the countryside. On top of this the brigades have been called upon to perform heroic deeds during the frequent earthquakes.

Unlike most fire companies the Chilean one has many mascots. These mascots are not dogs as you might imagine. They are small boys who will one day be full-fledged members of the volunteer system. Wearing uniforms like those of their big brothers and fathers, they learn the ways of putting out fires by watching and helping.

Acknowledgements

I am grateful for the information furnished by:
A. B. Advani, Chief Fire Officer, Delhi Fire Service Headquarters, New Delhi, India
H. F. Anguita, Superintendent of Chilean Firemen, Santiago, Chile
M. Arnaud, Chief of the Battalion, Chief of the Bureau, Paris, France
Laabib Badawy, Director, Public Relations Dept., Ministry of the Interior, Cairo, U.A.R.
J. Baekgaard, Copenhagen Fire Service, Copenhagen, Denmark
Nicholas Bouradas, Chief, Athens Fire Department, Athens, Greece
James W. Carter, Vice Consul, American Embassy, Santiago, Chile
Frank G. E. Coakwell, Chief, Toronto Fire Dept., Toronto, Canada
F. W. Delve, Chief Officer, London Fire Brigade, London, England.
Alfonso Grez, Consul General of Chile, New York City, New York
Istanbul Fire Department, Istanbul, Turkey
Enrico Massocco, General Director of the Fire Prevention Services, Ministry of the Interior, Rome, Italy.
C. E. Meek, Hon., Battalion Chief, Library of the Fire Dept., New York City, N.Y.
John A. McVicker, Consul, American Embassy, Moscow, U.S.S.R.
Giuseppe Migliore, The General Director of Civilian Protection and Fire Prevention Services, Ministry of the Interior, Rome, Italy
Frederick E. Myers, Consul, American Embassy, Cairo, U.A.R.
Takashi Onda, Asst. Information Officer, Consulate General of Japan, Tokyo, Japan
M. G. Pradhan, Commandant, National Fire Service College, Nagpur, India
President of the German Fire Defenses, Rootweil-Zimmern, Germany
P. A. Riedel, Commandant of the Amsterdam Fire Brigade, Amsterdam, Netherlands
Eulogio Samio, Fire Chief, Manila Fire Dept., Manila, Philippine Islands
Knud Skrivergaard, Asst. Press Attaché, Danish Information Office, Copenhagen, Denmark
Major General Pan Tun-Yi, Chief, Taipan Municipal Police Headquarters, Taipei, Taiwan, China
John R. Wood, Consul, American Embassy, Paris, France

Bibliography

Holzman, Robert S. — *The Romance of Firefighting*. New York Harper Brothers, 1956.

Kenlon, John — *Fires and Firefighters*. New York, George H. Doran, 1913.

Kermayr, Hans G. — *Der Goldene Helm*. Muchen, Pohl & Co., 1956.

Levine, Irving R. — *Main Street, Italy*. Garden City, N. Y. Doubleday & Co., 1963.

Norton, Howard — *Only in Russia*. Princeton, N. J., D. Van Nostrand Co., Inc. 1961.

Stevens, Leslie C. — *Russian Assignment*. Boston, Little, Brown & Co., 1953.

Pronunciation Guide to Foreign Words

AQUARI	ah-*kwah*-ree
CASTRA	*kast*-rah
CENTURION	sen-*tour*-ee-ahn
CHU JUNG	Choo Yung
DAIMYO	*dye*-mee-yoh
FATIMITES	*Fat*-ee-mites
HEPHAESTUS	Hee-*fess*-tus
KNOSSUS	*Nahs*-us
MACHINIKESHI	Mah-che-nee-keh-shee
NOCTURNES	nock-*tour*-ness
PREFECTUS	pre-*fect*-us
PTOLEMIES	*Toll*-eh-mees
QUESTIONARIUS	*kwest*-eh-ahn-air-eh-us
SAMURAI	*Sah*-moo-rye
SAPEUR-POMPIERS	seh-pur-*pomp*-ee-ay
SHOBO DEZOME-SHIKI	Sho-boh Deh-zoh-meh Shee-kee
SIPHONES	see-*fon*-ess
THULE	*Too*-le
VIGILUM	vee-*gee*-loom

Index

Italy, firefighters of, 30-33

Jamestown settlement fire (1608), 11
Japan, 41; firefighters of, 52-55

Kanto earthquake and fire, 55
Knossus fire (1400 B.C.), 38

Leningrad (St. Petersburg), first fire company in, 41
London: fire in (1666), 16; World War II blitz raids on, 17

Machinikeshi, in Japan, 54
Manila fire (1964), 57
Meireki fire (1657), 55
Metz, Carl, 23
Milan, fire brigade in, 33
Moscow: fires in, including Great Fire of 1812, 42; Volunteer Firemen's Association in, 43
Mount Holly, N.J., firehouse in, 11

Natron, powdered, for firefighting, 23
Nero, 32
Netherlands, firefighters of, 25-26
New Brunswick fire (1825), 13
New York City, firefighters of, 9, 11
Nocturnes, in ancient Rome, 31
Nuns of Ursberg, Germany, as firefighters, 23-24

Paris: fire bombs dropped on, during World War II, 29; fire watchers of, in 13th century, 28
Peshtigo fire (1871), 11
Philadelphia, colonial firefighters of, 9
Philippines, firefighters of, 56-57
"Powder" fire engine, first, 23
Prefectus vigilum, in ancient Rome, 32

Ptolemies, 45

Questionarius, in ancient Rome, 32

Rampur, National Fire Service College at, 47
Red Cross Society, of Denmark, 20
Rome: ancient, 6, 31, 32; modern, 33
Royal Palace, in Denmark, fire in (1884), 20
Russia, 33; firefighters of, 40-43

Salvage Corps, in Denmark, 21
Samurai Fire Brigade, 53-54
Santiago fire (1863), 59
Sapeurs-Pompiers, of France, 29
Satellite engine, 11
Shanghai, International Fire Brigade in, 50
Shobo Dezone-Shiki ceremony, 55
Siphones, in ancient Rome, 31
Smyrna, fire insurance brigade in, 35
Soap suds, for firefighting, 23
Superpumper engine, 11
Supertender engine, 11

Thule Air Base, fire department at, 21
Tokyo, firefighters of, 55
Toronto, first paid fire department in, 13
Turkey, firefighters of, 34-36, 41

United States, firefighters of, 8-11

Vatican Firefighters, 32
Venice, fireboats in, 33

Washington, George, 9
Water-pump, invented by Hiron, 45
Water tower truck, 11